# EVOLUTION AND CHARACTER

## (1908)

## BY

## ALFRED RUSSEL WALLACE

British Library Cataloguing-in-Publication Data
A catalogue record for this book is available from the
British Library

# Alfred Russel Wallace

Alfred Russel Wallace was born on 8$^{th}$ January 1823 in the village of Llanbadoc, in Monmouthshire, Wales.

At the age of five, Wallace's family moved to Hertford where he later enrolled at Hertford Grammar School. He was educated there until financial difficulties forced his family to withdraw him in 1836. He then boarded with his older brother John before becoming an apprentice to his eldest brother, William, a surveyor. He worked for William for six years until the business declined due to difficult economic conditions.

After a brief period of unemployment, he was hired as a master at the Collegiate School in Leicester to teach drawing, map-making, and surveying. During this time he met the entomologist Henry Bates who inspired Wallace to begin collecting insects. He and bates continued exchanging letters after Wallace left teaching to pursue his surveying career. They corresponded on prominent works of the time such as Charles Darwin's *The Voyage of the Beagle* (1839) and Robert Chamber's *Vestiges of the Natural History of Creation* (1844).

Wallace was inspired by the travelling naturalists of the day and decided to begin his exploration career collecting specimens in the Amazon rainforest. He explored the Rio Negra for four years, making notes on the peoples and languages he encountered as well as the geography, flora, and fauna. On his return voyage his ship, Helen, caught fire and he and the crew were stranded for ten days before being picked up by the Jordeson, a brig travelling from Cuba to London. All of his specimens aboard Helen had been lost.

After a brief stay in England he embarked on a journey to the Malay Archipelago (now Singapore, Malaysia, and Indonesia). During this eight year period he collected more than 126,000 specimens, several thousand of which represented new species to science. While travelling, Wallace refined his thoughts about evolution and in 1858 he outlined his theory of natural selection in an article he sent to Charles Darwin. This was published in the same year along with Darwin's own theory. Wallace eventually published an account of his travels *The Malay Archipelago* in 1869, and it became one of the most popular books of scientific exploration in the 19$^{th}$ century.

Upon his return to England, in 1862, Wallace became a staunch defender of Darwin's landmark work *On the Origin of Species* (1859). He wrote responses to those critical of the theory of natural selection, including 'Remarks on the Rev. S. Haughton's Paper on the Bee's Cell, And on the Origin of Species' (1863) and 'Creation by Law' (1867). The former of these was particularly pleasing to Darwin. Wallace also published important papers such as 'The Origin of Human Races and the Antiquity of Man Deduced from the Theory of 'Natural Selection" (1864) and books, including the much cited *Darwinism* (1889).

Wallace made a huge contribution to the natural sciences and he will continue to be remembered as one of the key figures in the development of evolutionary theory.

Wallace died on 7th November 1913 at the age of 90. He is buried in a small cemetery at Broadstone, Dorset, England.

# EVOLUTION AND CHARACTER

## INTRODUCTORY REMARKS.

In the very brief discussion of a great subject here attempted, I limit myself to a strict application of the modern doctrine of Organic Evolution to certain definite inquiries as to the probable development of human faculty or character. In doing so I first endeavour to define what is meant by evolution in general and by organic evolution in particular, and then proceed to show what are the essential agencies or processes by means of which the latter carries on its work. Dealing next with our more special subject, I inquire into the supposed differences between the minds of savages and those of civilised man, and also into those between our human nature to-day and that which existed in the earliest historic or pre-historic ages of which we have any records. Having thus arrived at certain probable conclusions, I proceed to suggest the conditions and agencies which are alone adequate to bring about a continuous advance in the average character of man during future ages.

# WHAT IS EVOLUTION?

The term Evolution, though now so generally used, is yet often misunderstood. It is supposed by many, perhaps by the majority of non-scientific persons, to indicate a great scientific theory which is applicable to and explains all the phenomena of the universe. But this is, very largely, an erroneous view. It is true that by many of its advocates it is held to be universally applicable, yet it has, so far, only given us a fairly complete explanation in certain departments of nature, and even in these it never carries us back to the beginnings of things; while over some of the broadest fields of scientific research it has been almost entirely inoperative.

Its essential idea is that of the continuity of all the phenomena of nature--that everything we see on the earth or in the spaces around us is not permanent, but has arisen out of something that preceded it. It is thus opposed to the old, and to some extent still prevalent, idea of creation--that things as we now see them have existed from some remote but definite epoch when they came into existence by the act or fiat of a supreme power--the great First Cause. Evolution is thus a general statement that everything is, and always has been, slowly changing under the influences of natural laws and processes; but, except in a few cases, it cannot give a precise account of the methods and causes of the changes, still less can it carry us back to any beginning of the universe.

It thus formulates a general process, but is unable to give us any full explanation of that process.

# SLOW GROWTH OF THE IDEA OF EVOLUTION.

Although the philosophers of Greece had vague ideas of evolution, which were elaborately worked out by the Roman poet Lucretius in his great poem "On the Nature of Things," yet their views obtained no general acceptance until our own era, owing mainly to the positive statements as to the creation of the universe in the "Old Testament," and the very general acceptance of that record as the Word of God. Even the very surface of the earth was held to be unchangeable, as implied in the term "the everlasting hills"; while less than fifty years ago so great a writer as George Borrow could speak of a waterfall as being in all details as it was "since the day of creation, and will probably remain to the day of judgment."

The modern view of continuous change by natural forces was first applied to the surface and structure of the earth, by more or less careful observation of the facts and phenomena included in the modern science of geology. It began with a few acute observers in the seventeenth century, among whom were Leibnitz and Hooke, followed by numerous

Italian writers in the eighteenth century, together with the Germans Pallas and Werner; while our own countryman, Hutton, for the first time laid down the great principle of modern earth-study, that we can only understand the past by a careful study of the various changes we now perceive to be in progress. This great principle was afterwards most skilfully applied by Sir Charles Lyell, who devoted a long life to the production and continuous amplification of his monumental work, *The Principles of Geology*. The idea of evolution was thus applied in detail to one of the greatest and most complex departments of human knowledge.

In the two other great sciences dealing with the constitution of the inorganic world--chemistry and astronomy--progress was for a long time necessarily limited to the study of facts and phenomena, with the laws to which those phenomena are immediately due; and the conclusions arrived at pointed rather to stability and permanence than to that progressive and unceasing change that is the keynote of evolution. The great mathematicians, who at the beginning of the nineteenth century worked at the motions and disturbances in our solar system in accordance with the law of gravitation, came to the conclusion that the system was a stable one, that all irregularities were slight and temporary, and that the planets and their satellites were so arranged that their present positions and motions might continue for ever without any destructive changes.

In like manner it was for long a fundamental doctrine of chemistry that the elements were fixed and unchangeable, and the belief of the alchemists that other substances might be converted into gold was held to be as baseless and as unscientific a dream as the idea that matter itself was destructible.

But in our own day, and largely by the work of men still living, all these assumptions of indestructibility and permanence have been rudely shaken or altogether given up as in their turn unscientific. Through the development of what may be well termed the two modern sciences--electricity and spectrum analysis--together with the systematic study of the long-neglected phenomena of meteors and meteor systems, quite new conceptions have been reached as to the constitution of the universe, ascending, on the one hand, to the nature and origin of the myriad stars and suns and nebulæ which constitute our universe, and descending on the other to the nature, the constitution, and even the instability of matter itself, as indicated by the strange and almost incredible phenomena presented by radio-active substances.

By these various advances in many directions we have attained the certainty that the great principle of evolution pervades the entire realm of nature, from the faint specks of star-dust on the farthest limits of our stellar universe, down to what were once supposed to be the indestructible atoms of matter itself, now proved to be complex systems of electric force-points, subject to disturbances and even to absolute disintegration. We thus seem able dimly to comprehend on the one hand the evolution of matter itself, with its marvellous properties, which enable it to become manifest in the myriad forms made known to us by the chemist or existing in the vast laboratory of nature; on the other hand, the evolution of this matter into the inconceivably vast and complex stellar universe. Everywhere we behold a state of flux, of development, and also, apparently, of decay. Every increase of knowledge seems to imply that the material universe is a vast organism which must have had a beginning and will have an end--which was born and will die. The dissipation of energy and the disintegration of matter alike render this conclusion logically certain.

## ORGANIC EVOLUTION.

The preceding remarks apply only to what may be termed inorganic or physical evolution, which necessarily preceded and prepared the way for the evolution of the organic world--an evolution which is utterly unlike anything which preceded it and which has culminated in the production of man--the one being who is able, to some extent, to comprehend the universe of which he forms a part, to penetrate to its remotest confines, to study its laws and to speculate on its nature, its origin, and its destiny. Hence we may naturally conclude that the final law and purpose of the whole universe was the development of so marvellous a being who has been deemed to be "a little lower than the angels" and "in apprehension like a god."[1]

## ORGANIC AND INORGANIC EVOLUTION COMPARED.

It will be well to note here the fundamental difference between organic and inorganic evolution, a difference so great and so radical that it is somewhat misleading to use the same term to describe them both. The changes that occur in the inorganic matter of the universe are of three kinds. (1) The changes of eternal form, and to some extent of internal structure, caused by the disintegration and aggregation of masses of matter, as in the formation of most rocks and in the successive modifications of the earth's surface. (2) The changes in the state of matter--solid, liquid, and gaseous-- dependent on the amount of heat received from without or generated within cosmical bodies. (3) Chemical changes, by which the very constitution of matter becomes so modified as to give each new compound a special character and properties which it usually retains unchanged for indefinite periods. The various kinds of matter produced by these changes seem to be permanent so long as the conditions remain the same, but, except in the case of crystals, they are of no definite shape or size. By changes of conditions they become to some extent interchangeable, but, except when subjected to these changes, they remain inert, or subject to excessively slow processes of degradation or decomposition, whence the common term dead matter.

Organic evolution, on the other hand, leads to the production of highly complex individual entities of definite forms, each in a state of constant internal movement, each permeated by liquids and gases by means of which they assimilate new matter from the outer world, change that matter into new forms that occur nowhere else in nature, and are enabled to carry on the mysterious processes of growth and reproduction. Each of these individuals, beginning with minute cells in the body of a parent, passes through a regular cycle of growth, maturity, and decay, culminating in what we term death, when all its regular internal motions cease, it becomes disintegrated by the agency of lower organisms, and finally helps to build up other forms of life. Each reproduces its kind almost identical in all respects with the parent, thus forming a cycle which was long believed to be perpetual and unchanging, the species of animals and plants being held to be fixed entities produced by some act of creative power. The great and unique phenomenon of the organic world is reproductive growth by the absorption and transformation of inorganic and organic matter, and the building up again and again of a highly complex organism from a single cell. It is this wonderful process that we term Life, whether manifested in animals which possess sensation and in their higher forms consciousness, or in plants which there is no reason to believe possess these faculties.

## THE LAW OF ORGANIC EVOLUTION.

At about the middle of the last century so great a man of science as Sir John Herschell spoke of the mode of origin of the various species of animals and plants as being the "mystery of mysteries"; for although many writers had discarded "special creation" and had expressed a belief in evolution through the normal process of generation, yet no one had shown *how* the various species and genera had been produced, or by what means the wonderful adaptation of each to its special conditions of existence had been brought about. The problem was, however, solved by Charles Darwin almost coincidently with the other great advances in the domain of inorganic nature already alluded to.

The fundamental law which he discovered, and was the first to develop in all its far-reaching results, is that of "Natural Selection," or the "Survival of the Fittest." This great law of nature is the result of a group of well-known and universal facts--(1) the enormous powers of increase of all organisms, an increase so great that any one of them, if left alone in an unoccupied continent, would fill it to overflowing in a few years or centuries. As, however, the whole earth is already occupied, this simultaneous increase of all the thousands of species in every country produces a "struggle for existence," there being no room for the new-comers under average conditions till the parents have ceased to exist; and as all the higher animals (and plants) live several years, breeding every year, it is evident that on the average all their progeny must die from various causes before, or shortly after, they arrive at maturity except *one pair* to replace the two parents.

Now comes the question, is the destruction of the superfluous thousands a matter of pure chance, or is there any cause why certain individuals should survive the rest? If the offspring were always identical copies of the parents, not only in external form, but in every internal character and quality, in health, in acuteness of the senses, in activity, and in all the mental powers and faculties, then we should be obliged to impute to chance alone the destruction of ninety-nine while one survived. But we know there is no such similarity. In every large family of children considerable diversities occur as the rule rather than as the exception. In every litter of kittens or of puppies there are similar differences; while it has been through the selection of some of these varieties and the rejection of others that all our fine breeds of horses, cattle, and sheep have been produced, as well as all our fancy pigeons and poultry. And we now know that exactly the same thing occurs among animals and plants. By collecting and examining hundreds or thousands of individuals in one district and during one season, an amount of variation is found to exist much greater than anything that even Darwin expected. He sometimes spoke of nature having to wait for "favourable variations"; of natural selection being powerless unless "favourable variations" occurred when required. But these doubts and hesitations are utterly needless. There are *always* favourable variations in every direction, and in ample quantity. Take any measurable character you please, and in 50 or 100 or 1,000 individuals about one-third or one-fourth will be considerably above or below the mean, so much so as to be distinctly visible to us, while about one-third or one-half are so near the average of the whole that on a cursory inspection we should say they were all alike. But as, every year, only from one-tenth to one-hundredth of the young of a given species can survive, there is always an ample supply of "favourable variations." We must remember also that nature does not select, as we are often obliged to do, by the size or form of any limb, part, or organ, but by the resultant *qualities*, and we know that these qualities vary as much as the parts of the body we can measure or estimate. Horses, from the same or equally good parents, vary in speed and in endurance; dogs in acuteness of smell or of sight; sheep in the thickness of their wool; cows in their milk-giving capacity. In all these cases there is a considerable amount of better or of worse, offering ample "favourable variations." During an exceptionally severe winter only the swiftest and most enduring wolves survive, the rest perishing of cold and hunger. In a prolonged drought it is only the tallest giraffes that find food enough to support life; and thus, by a periodical weeding-out of all but the very best--the fittest to survive under these unfavourable conditions--the standard of efficiency in each species is preserved by the rigid destruction of the less fit. It must always be remembered that, although the *average* population of each species varies very little during long periods, yet there may be considerable fluctuations annually. Some seasons will favour one species, some another; we then notice the abundance of certain birds or insects, generally followed a year or two later by a corresponding scarcity, keeping up the balance of the various forms of life in generally uniform proportions so long as the natural conditions, or "environment," continue to be the same or nearly the same.

## THE ORIGIN OF SPECIES.

To anyone who has thoroughly grasped the extent and

universality of variation within the limits of every common or widely-spread species, it will be at once evident that the very same causes which preserve each species in exact adjustment to its environment, will also, when that environment changes in any direction, enable it to become automatically adjusted to the new conditions. This must be the case, because all alterations in environment are necessarily very gradual. Changes of climate require thousands of years before they attain an amount greater than occurs during the ordinary seasonal or periodical changes to which all animals and plants are already adjusted. If a new enemy enters a country, it requires a number of years, perhaps centuries, before it can become itself adapted to the new conditions and increase to an extent sufficient to endanger any considerable number of other species. Some of the weaker kinds, and a proportion of the young of the stronger and more numerous, will no doubt fall a prey; but this will itself lead to the adaptation and improvement of the remainder, since those that escape will inevitably be those who are best fitted, either by swiftness, or strength, or cunning, or by possessing some special coloration or peculiar habit that conceals them from the enemy. The very danger itself leads to such a gradual modification of the sufferers as to enable those that escape to become the progenitors of a race better fitted to cope with the new danger.

Even more certainly automatic are the effects of climatic

or food changes. To some these changes will be injurious, to some indifferent, to a few perhaps beneficial. The former will be weakened and thus fall a prey to the other causes of destruction, while the latter, being actually improved in health, will always furnish a large proportion of those that survive and become the parents of new races.

It is thus evident that those species which are already best adapted to the environment in a large country, that have a wide geographical range and exist in very great numbers, will be those that, by furnishing a large amount of variation every year, will be best able to become rapidly modified into an exact adaptation to any new conditions that may arise. These were termed by Darwin dominant species; and, owing to their general variability and enormous numbers, it is these that usually become the parents of a whole series of diverging species when exposed to a variety of new conditions in different parts of their area.

But besides these dominant forms other species exist which are rare or local, sometimes inhabiting only a very limited area, or being specially adapted to a very restricted mode of life--as when a caterpillar feeds on one species of plant only, and that not a common one. These are the species which are either dying out from want of power to compete with other species, or are so specially adapted to a limited environment that a comparatively slight change leads to their extinction. It is these which become extinct without leaving modified descendants, as seen in the countless genera and families which have successively died out in each geological epoch. The dominant groups, on the other hand, can often be traced far back in geological time, as in the well-known case of the horse tribe, the cats, and some others; or they are those which, though of comparatively recent origin--as the deer and the antelopes--are so well adapted to existing conditions that they have spread over wide areas of the earth's surface with numerous specific forms adapted to local peculiarities of environment.

## UTILITY THE KEY-NOTE OF ORGANIC EVOLUTION.

If the facts of nature now briefly sketched are clearly apprehended, there result two propositions of the highest importance. These are: (1) That whatever the amount of variability of a species, no general modification of it will occur so long as the environment remains unchanged; and (2) that when a permanent change (not a mere temporary fluctuation) of the environment occurs--whether of climate, of extension or elevation of land, of diminished food-supply, or of new competitors, or of new enemies--then, and then only, will various specific forms become modified, *so as to adapt them more completely to the new conditions of existence.*

21

It is easy to see that all the kinds of changes above indicated are so connected that they will inevitably occur together, though in various degrees. Changes of climate or in the area or elevation of the land will cause changes in the vegetation; this will afford more or less food to various kinds of animals, and these animals will again be preyed upon by other animals. At the same time, most of these animals will need adaptation to the climate itself, whether hotter or colder, wetter or dryer; and each will also have new competitors and more or fewer enemies. There will thus be action and reaction of the most complex kind, and if the change of environment is great in amount and extends over a wide area, then, when it is completed, almost all the species that formerly existed will be found to have become more or less changed in form, structure, and habits so as to constitute new species, and in some cases new genera. This amount of change has again and again actually occurred as shown by the extinct animals and plants preserved in the rocks of the various sub-divisions of the Tertiary Period. On the other hand, we find proofs of periods of stability in the fact that certain deposits which, from their extent and thickness must have required a long period of time for their formation, yet contain from top to bottom an almost identical series of species.

The various modifications of form, structure, or colour thus produced constitute the "specific character" of each new species, distinguishing it both from its parent species (which will usually have become extinct) and from all its near allies, and each of these characters must have been at the time they were fixed by continued selection, *useful* to the species. This has been, and still is, denied by many naturalists, mainly because they cannot see or imagine any use in many characters; but they have never succeeded in discovering any cause other than utility adequate to produce new characters which shall be present in all the individuals of a species and strictly confined to them. In order to be developed through natural selection a particular variation must not only be *useful*, but must, at least occasionally, be of such importance as to lead to the saving of life, or, to use Professor Lloyd-Morgan's suggestive term, be of "survival-value." This subject has been somewhat fully discussed in Chapters V. to X. of my *Darwinism*, supplemented and enforced by the five chapters on the "Theory of Evolution" (in my *Studies*, Vol. I.), and to these works I must refer my readers for fuller information. There is, however, one other preliminary question of special importance as regards the subject of the present essay, that must be briefly discussed.

# THE INHERITANCE OF ACQUIRED CHARACTERS.

A problem of the highest importance as regards the whole theory of organic evolution, and especially in its varied applications to man's nature and advancement both individually and socially, is to determine the limits of heredity. The first great writer who put forth a detailed theory of the method of organic evolution was Lamarck, who believed that the chief cause of the modification of species was desire and effort, leading to the use of certain organs and parts to their fullest capacity; that such use strengthened and enlarged such organs (as it undoubtedly does); and that this increased development was transmitted to the offspring. In this way he thought that all the adaptations of animals to their mode of life--the strength of the lion, the speed of the antelope, the long neck of the giraffe, and all other such characters, had been acquired.

But his facts and arguments, though highly ingenious, made little impression, mainly because naturalists perceived that his theory was only applicable to a very small portion of the adaptations which needed explanation. As examples, no exercise of the will or of the muscles could produce those wonderful harmonies of colour which serve to conceal so many herbivorous creatures from their enemies and enable so many of the rapacious kinds to approach their prey. Neither could any similar action of body or mind lead to the growth of the wonderful shells of the mollusca, the bony armour of the tortoises, or the poison-glands of snakes or of stinging-insects. Still less can such causes have been effective in the production of countless adaptations in the whole structure of plants, and especially of their flowers and fruits.

But although Lamarck's theory was seen to be utterly inadequate, its fundamental assumption--that the effects of the use and disuse of organs were transmitted to the offspring--seemed so probable that it was generally accepted without any critical examination, and was even thought by Darwin to be a useful adjunct to his own theory. This was partly due to the fact that none of the early naturalists paid much attention to *variation*, which was only a source of trouble to them in their efforts to define "true species." But with the interest excited by Darwin's works the study of varieties was seen to be of the first importance, since they are the materials out of which new species are formed. It was then very soon found that variations are much more abundant, much larger in amount, and much more varied in character than was supposed, and that, together with the ever-growing proofs of the extreme rigour of natural selection, a sufficient explanation of the origin of all species was attained.

Then followed the investigations of Galton and Weismann, showing that there was no valid evidence of the transmission of the modifications of individuals due to use or disuse, or to climate, food, or other external agencies, while the elaborate researches of the latter into the earliest processes of reproduction resulting in his illuminating theory of the "continuity of the germ-plasm," gave what was probably the death-blow to Lamarck's fundamental assumption.

During the last decade evidence has been accumulating to prove that, among the higher animals at all events, it is only the inborn characters and tendencies--whether physical or mental--that have any part in producing the varying characters of the offspring, and at the present time it may be said that almost all the chief biological thinkers and investigators hold this view, including Professors Ray-Lankester, Lloyd-Morgan, E. B. Poulton, and Sir W. T. Thiselton-Dyer. The latest and one of the acutest students of this and allied problems--Mr. G. Archdall Reid--in his remarkable work, *The Principles of Heredity*, sums up the whole case against the heredity of acquired characters with great lucidity, and shows that instead of being, as Darwin and others thought it might be, an aid to natural selection in bringing about adaptation to new conditions, it is really in the great majority of cases antagonistic to it, while in some it would actually neutralise it altogether. Those who are interested in this problem should study Mr. Reid's work, or the smaller but equally conclusive treatise by Mr. W. Platt Ball, *Are the Effects of Use and Disuse Inherited?* in which the evidence on both sides is fairly given and the answer shown to be unmistakably in the negative.

# HEREDITY IS THE SAME FOR MIND AS FOR BODY.

Before leaving this part of our subject it may be as well to state that, broadly speaking, every fact and argument here given applies to the mental as well as to the bodily organs, to the intellectual as well as to the physical powers of animals including man. Galton has proved that genius, like physical qualities, is hereditary, and all other mental faculties are equally so. Great genius, like gigantic stature or enormous strength, is rare, and men in general approximate to the average in mind as well as in body. The only important difference seems to be that the mental faculties vary to a greater extent than the bodily organs. Newton and Cayley in mathematics, Shakespeare and Shelley in poetry, rise higher above the average man in intellect than do the equally rare giants in stature. This feature will be referred to later. Again, all the evidence goes to show that, though native or inborn faculties are hereditary, mental acquisitions or the results of education or experience are *not* transmitted any more than those of the bodily organs. If such acquisitions were transmitted, we should expect the younger sons of great men to possess their fathers' abilities to a greater degree than the elder sons, since they should inherit some of the ability due to the practice of the father's art or profession for a longer period; but no such constant difference has ever been detected. Still more conclusive is the fact that the children of English, French, or German parents, whose ancestors for many generations have spoken each their own tongue, do not show any exceptional power of learning their own rather than another language; and this applies in cases where the speech is most diverse, as when a European infant is reared among Hindoos or Red Indians. These various considerations render it almost certain that the phenomena of heredity and variation are fundamentally the same in the mental as in the physical departments of human nature. This being the case, we must assume that character (which is really the aggregate of the intellectual and moral faculties), in order to be progressively developed, must be acted upon by some form of natural selection. We have seen, however, that this power only acts--can only act--by the survival of individuals which possess the more *useful* developments. It follows that those special faculties which build up character can only be preserved and increased inasmuch as they are of use to the individual or to the race, and this *utility* must be of such a nature as in times of stress or danger to be of *life-preserving value*. We must, therefore, proceed to inquire in what way and to what extent character has been, or is being, modified or advanced.

# THE EVOLUTION OF CHARACTER IN THE INDIVIDUAL AND IN THE RACE.

Every evolutionist now believes that man has arisen from the lower animals by a process of modification, in the same way as any species of animal has arisen from its ancestral forms. He is the culminating point of the whole vast fabric of the organic world. If he has not so arisen, but is the product of other unknown forces guided by infinite power, then the slow development of the infinitely varied forms of nature that preceded his advent appears to be unmeaning. But if there is any purpose in the universe, if nature and man are not the chance products of primeval forces, we must conclude that the process by which man has actually been developed is the best, perhaps the only possible, mode of producing him. From this point of view everything is harmonious and intelligible. The end to be attained, required, and justified, the countless ages of preparation, of which we obtain some imperfect knowledge in the geological record. The varied forms of vegetable and animal life which filled the earth when man first appeared afforded him the means of life, not in one part only, but over the whole terrestrial surface. As he advanced in knowledge and increased in population, an ever-increasing proportion of plants and animals became of use to him, first for food, then for weapons, for clothing, for houses, for utensils; and later on for comfort and luxury, as aids to his mental development or to charm him by their beauty. The marvellous phenomena of nature, from the glittering hosts of heaven to the exquisite panorama of the seasons on the earth, awoke in him the desire for knowledge; and, as time went on, ever more and more of the secrets of nature were revealed to him, ever more and more of her powers were utilised, an ever-increasing proportion of the animal and vegetable and mineral worlds became subservient to his needs, or gratified his intellectual or æsthetic or moral faculties. Yet more, if there is a purpose in the universe, if the organic world came into existence in order that man might exist, then we must also recognise purpose in that infinite *variety* of nature, whether animate or inanimate, which has furnished such an inexhaustible supply of everything necessary for his life and happiness, and for the progressive development of his intellectual and moral nature. We can believe (and not be afraid to acknowledge our belief) that the dog and the cat, the sheep and the cow, the horse and the ass, the fowl and the pigeon, the throstle and the nightingale, the orange and the apple, the strawberry and the vine, wheat and maize, pine-tree and oak, and all the myriad luscious fruits and fragrant flowers and glorious blossoms, and infinitely varied mineral and chemical products--all alike exist as parts of the great design of human development.

And here again we obtain further indications of purpose in the very method of organic as distinct from that of inorganic evolution. The two great distinctive features of living substance--enormous powers of increase together with gradual but almost unlimited variability (features that are absent from the entire inorganic universe)--necessarily lead to the rapid spread of life over the whole area that is not absolutely unfitted for it, and at the same time give rise to an ever-increasing variety of forms and complexity of structure, in adaptation to the ever-changing conditions of the earth's surface. Thus the whole earth and ocean have become filled with continually varying and progressing forms of life, so that when the cosmic process culminated in man, with faculties and aspirations calculated to utilise and appreciate them, it also culminated in those highest developments of the animal and vegetable worlds which we have briefly enumerated and which certainly never existed together in equal variety and beauty, at any earlier period of the earth's history.

## IS THERE ANY EVIDENCE OF PROGRESSIVE IMPROVEMENT IN CHARACTER?

But though it is admitted that man has arisen from a lower animal form, we have still to inquire whether his whole intellectual, æsthetic, and moral nature has been produced by the action of the very same laws and processes as have led to the development of animal forms and animal natures. Does variation and survival of the fittest explain man's mind as well as his body? Does he differ from the lower animals in degree only, or is there an essential difference in his mental nature?

It is clear that from brute to man there has been a great advance. This is universally admitted. But that there has been any very great advance from the earliest men of whom we have any records and ourselves, is by no means generally admitted and certainly cannot be proved.

I have myself shown that the great first step that caused man to rise above his fellow animals was that amount of mental superiority that enabled him to obtain some command over nature. After showing how each animal form could only preserve its existence in a changing universe by corresponding changes in bodily structure or in the lower mental faculties, I go on to describe what occurred in the case of man:--

"At length, however, there came into existence a being in whom the subtle force we term *mind*, became of more importance than his mere bodily structure. Though with a naked and unprotected body, *this* gave him clothing against the varying inclemencies of the seasons. Though unable to compete with the deer in swiftness or with the wild bull in strength, *this* gave him weapons with which to capture or overcome both. Though less capable than most other animals of living on the herbs and fruits that unaided nature supplies, this wonderful faculty taught him to govern and direct nature to his own benefit, and make her produce food for him when and where he needed. From the moment when the first skin was used for a covering, when the first rude spear was formed to assist in the chase, when fire was first used to cook his food, when the first seed was sown or shoot planted, a grand revolution was effected in nature--a revolution which in all the previous ages of the earth's history had had no parallel--for a being had arisen who was no longer necessarily subject to physical change with the changing universe, a being who was in some degree superior to nature, inasmuch as he knew how to control and regulate her action, and could keep himself in harmony with her, not by a change in body, but by an advance in mind."[2]

Now this passage, first published in 1864, seems to me to indicate the essential superiority of man over the lower animals, a superiority which was perhaps as great fundamentally in palæolithic or eolithic man as it is now. All that we have done since, all the triumphs of our civilisation and of our science, have arisen by slow, very slow, progressive steps, each one only a little in advance of what had been done before, and none of them perhaps so difficult, so clearly showing superiority of intellect, as those marvellous first steps which proved that a new and a higher being had appeared on the earth.

Mr. Archdall Reid, in his work on Heredity already referred to (in a very suggestive chapter on "Racial Mental Differences") adopts the views of Buckle and John Stuart Mill, that by far the larger part of racial or national differences of character are *not* inherent, but are the product of the diverse and highly complex environments of each. This would include, of course, their past history, their religion, their education, their form of government, and the various habits and customs, language, legends and superstitions that have come down to them from a forgotten past. In comparing a savage with a civilised race, we must always remember that the amount of acquired and applied knowledge which we possess is no criterion of mental superiority on our side, or of inferiority on his. The average Zulu or Fijian may be very little lower mentally than the average Englishman; and it is, I think, quite certain that the average Briton, Saxon, Dane, and Norseman of a thousand years ago--the ancestral stocks of the present English race--were mentally our equals. For what power has been since at work to improve them? There has certainly been no special survival of the more intellectual and moral, but rather the reverse. As Galton points out, the celibacy of the Roman Church and the seclusion of thousands of the more refined women in abbeys and nunneries tended to brutalise the race.

To this we must add the destruction of thousands of psychics, many of them students and inventors, during the witchcraft mania, and the repression of thought and intellect by the Inquisition; and when we consider further that the effects of education and the arts are not hereditary, we shall be forced to the conclusion that we are to-day, in all probability, mentally and morally inferior to our semi-barbaric ancestors!

Looking back at the course of our history from the Saxon invasion to the end of the nineteenth century, what single cause can we allege for an advance in intellect and in moral nature? What selective agency of "survival value" has ever been at work to preserve the wise and good and to eliminate the stupid and the bad? And it must have certainly been a very powerful agency, acting in a very systematic manner, even to neutralise the effect of the powerful deteriorating agencies above referred to.

When we remember that the Romans and the Greeks looked down on *all* our ancestors as we look down on Kaffirs and Red Indians, we must not too hastily conclude that, because people are in the savage or barbarian state as regards knowledge and material civilisation, they are necessarily inferior intellectually or morally. I am inclined to believe that an unbiassed examination of the question would lead us to the conclusion which, as I understand him, is favoured by Mr. Archdall Reid, that there is no good evidence of any considerable improvement in man's average intellectual and moral status during the whole period of human history, nor any difference at all in that status corresponding with differences in material civilisation between civilised and savage races to-day. What differences actually exist are sufficiently accounted for by various selective agencies known to have been at work; while there is good reason to believe that some of the lowest savages to-day (perhaps all of them) are the deteriorated remnants of more civilised peoples.

## THE CHARACTER OF SAVAGES NOT NECESSARILY LOW.

If we turn to the facts actually known to us about early man, historic and prehistoric, they certainly point in the same direction. Whence came the wonderful outgrowth of art manifested by the Germans and Celts in their Gothic architecture, admirable alike in structure, in design, and in ornament, and which we, however much we pride ourselves on our science, cannot approach in either originality or beauty? Going further back, the Roman architects, sculptors, poets, and literary men were fully our equals. Still earlier, the Greeks were our equals, if not superior in art, in literature and in philosophy. The Aryans of Northern India were equally advanced, and their wonderful epic--the Maha-Bharata--introduces us to a people who were probably both in intellect and in morality no whit inferior to ourselves.

Further back still, in ancient Egypt, we find in the Great Pyramid a structure which is the oldest in the world, and in many respects the most remarkable. In its geometrical proportions, its orientation, and its marvellous accuracy of construction, it is in itself the record of a people who had already attained to a degree of high intellectual achievement. It was one of the most gigantic astronomical observatories ever erected by man, and it shows such astronomical and geometrical knowledge, such precision of structure, and such mechanical skill, as to imply long ages of previous civilisation, and an amount of intellect and love of knowledge fully equal to that of the greatest mathematicians, astronomers, and engineers of our day.

And if from the purely intellectual we turn to the domain of conduct and of ethical standards, we encounter facts which also lead us to the same conclusion. If we compare the two greatest ethical teachers of our age with their earliest prototypes whose works have been preserved, it is impossible to maintain that there has been any real advance in their respective characters. Tolstoy can hardly be ranked as higher than Buddha, or Ruskin than Confucius; and as we cannot suppose the amount of variation of human faculty about a mean value to be very different now from what it was at that remote era, we must conclude that equality in the highest implies equality in the mean, and that human nature on the whole has not advanced in intellect or in moral standards during the last three thousand years, while the records of Egypt in both departments--those of science and of ethics--enable us to extend the same conclusion to a period some thousands of years earlier.

In reply to this argument, it will be urged that the period from these early civilisations to our own day is only a fragment of man's whole history, and that in the remains of neolithic, palæolithic, and eolithic man, we have certain proofs that his earliest condition was that of a low and brutal savage. But this is pure assumption, because the evidence at our command does not bear upon the question at issue. Material civilisation and inherent character are things which have no necessary connection. There is no inconsistency, no necessary contradiction, in the supposition that the men of the early stone age were our equals intellectually and morally. As Mr. Archdall Reid well argues, if a potential Newton or Darwin were occasionally born among savages, how could his faculties manifest themselves in that forbidding environment? With an imperfect language and limited notation, and having to maintain a constant struggle for existence against the forces of nature, and in combination with his fellows against wild beasts and human enemies, either of them might have made some one step in advance--might have invented some new weapon or constructed some improved trap. He must necessarily work on the lines of his fellows and with the materials to his hand. Perhaps in the rude drawings of animals on stone or tusk we have the work of a potential Land-seer; while the equal of our Watt or Kelvin might have initiated the polished stone axe or invented the bone needle. That a people without metals and without written language, who could therefore leave few imperishable remains, may yet possess an intellect and moral character fully equal (some observers think superior) to our own, is demonstrated in the case of the Samoans, and some other tribes of the Pacific, It is clear, therefore, that a low state of material civilisation is no indication whatever of inferiority of character.

## DIVERGENCE OF NATIONAL AND INDIVIDUAL CHARACTER.

But although every indication of history and of existing races of man negatives the idea of any general *advance* of character, a conclusion which is supported by the entire absence of any selective agency of "survival value," which could alone have led to such advance on the principles of organic evolution, yet there are undoubtedly *differences* of national character which it is not easy to account for. That, on the whole, the Celtic races are more idealistic, more joyous, and more excitable than the Germanic or Sclavonian, while a similar difference exists between the peoples of Southern and Northern Europe, seems to be generally admitted. Buckle, as has been already noted, explained the difference by the influence of the diverse environments, and Mr. Archdall Reid favours the same view, but there are many difficulties connected with such a theory. No doubt the best known Celtic races--the Bretons, the Welsh, the Cornish, and the Highlanders--have been long the inhabitants of mountainous districts to which they have been driven by the invasions of more warlike peoples; but, unless some form of selection comes into play, it is difficult to imagine why this should have changed the character of people who had presumably lived at some earlier period in less awe-inspiring lands.

But the great argument against this explanation is to be found, I think, in the diverse characters of two of the principal divisions of mankind--the Mongoloid and the Negroid. Here we see that great changes of natural environment have produced no corresponding modification of character, and *vice versâ*. Every reader of my *Malay Archipelago* will, I think, remember my description of the Ke Islanders (typical Papuans and Negroids) and my comparison of their behaviour with that of the Malays (equally typical Mongoloids), with whose character I was so well acquainted. Now, the fundamental features of the *characters* of these two great divisions of mankind maintain themselves wherever they are found, in every variety of aspect and of climate, extending over three-fourths of the globe. The Red Indians of America (true Mongoloids for the most part) have the same impassive, unexcitable character in the frigid, the tropical, and the temperate zones; whether they inhabit the forests or the plains, the great river valleys or the lofty plateaux; and the same may be said of the Old World branch from the Japanese and Chinese to the Kalmucks and Malays; and, throughout these vast diversities of natural environment it cannot be said that any minor diversities of character can be positively traced to local influences. In the case of the Malays and Papuans, we have the two races existing under almost identical circumstances in the vast equatorial forests extending from Sumatra to the Solomon Islands, often living in a very similar manner and in an almost identical stage of barbarism; yet it is in this very region that their distinctive mental characteristics are to be found to be at a maximum. For such a mental divergence as these two races present, I cannot myself see any possibility of an explanation through any selective agency of "survival value"; while the influence of environment is equally untenable, besides being in direct opposition to the now well-established principle of the non-heredity of acquired peculiarities.

In this undoubted difference of racial character, and perhaps to an even greater extent in that of national character, the mental divergence seems to exceed the physical. The former better corresponds to the amount of mental difference between different species, genera, or even families of animals than to those presented by mere varieties of a single species; and in this way we have an indication of a want of parallelism or of direct relationship of the mental and the physical characteristics of mankind, which may, perhaps, offer us a clue to this most complex and important problem. Among individuals we see the same phenomenon, though we have no means of accurately estimating it. The amount of divergence in the physical features of healthy and equally well-nourished and well-trained individuals in the same country is not very great. In stature, strength, speed of running, and acuteness of the senses, the divergence from the mean is rarely more than as two to three, and in the most extreme cases does not exceed two to one. But in the mental faculties, or in any special faculty, the divergence would be usually estimated at a much higher figure. There are thousands of mathematicians among us to-day whose capacities would certainly be estimated at five or six times greater than that of other thousands who can never understand comparatively simple arithmetical or geometrical problems, while the extreme cases of the highest mathematical genius and the lowest degree of arithmetical stupidity would be estimated as at least some such proportion as 100 to 1, if not much higher; and in every other department of human faculty--music, poetry, or eloquence--there is perhaps a nearly equal amount of divergence.

We may, I think, explain this circumstance by the consideration that, while the physical characteristics must have been rigidly selected during the earlier period of man's existence on the earth, through his constant struggle with the lower animals and against the forces of nature, and later on almost equally so in war or competition with other tribes or races, his higher mental faculties were seldom or never called into action, being of no direct use to him in the struggle for existence. While the former, therefore, became fixed within definite limits, the latter were free to vary in amount through the agency of some unknown laws or inherent capacities. The extraordinary thing is, that these higher faculties did not become atrophied by disuse as would physical characters under similar conditions; instead of which they appear to have persisted undiminished in power throughout all human history, ready under favourable conditions to blaze out in a Homer or a Socrates, a Pyramid designer or a Buddha, an Archimedes or a Shakespeare.

Some of the greatest upholders of the theory of natural selection admit that these higher faculties cannot have been developed through its agency. In an elaborate essay on *The Musical Sense in Man and Animals*, Weismann comes to the conclusion that the musical sense "is simply a by-product or accessory of the auditory organ," and that it is "a merely incidental production, and thus, in a certain sense, an unintended one." In another work (his lecture on

*Heredity*) he arrives at a similar conclusion with regard to all the higher activities of the mind in the following statement:--"In my opinion, talents do not appear to depend upon the improvement of any special mental quality by continual practice, but they are the expression, and to a certain extent the by-product, of the human mind, which is so highly developed in all directions." Huxley arrived at a somewhat similar conclusion, being reported by Mr. Wilfred Ward as saying:--"One thing which weighs with me against pessimism, and tells for a benevolent author of the universe, is, my enjoyment of scenery and of music. I do not see how they can have helped in the struggle for existence. They are gratuitous gifts."

## WHAT FACULTIES HAVE BEEN SELECTED AND IMPROVED?

But though there has, apparently, been no continuous advance in the higher intellectual and moral nature of man for want of any selective agency leading to such a result, this has not been the case with that portion of his faculties which he possesses in common with the lower animals. The family affections, and the social instincts, were essential to the safety of the clan or the tribe; courage and perseverance, cautiousness and decision, were valuable in hunting and in war; the inventive and constructive faculties were of value in the making of weapons and snares, clothing and houses, while foresight, and the love of animals, might lead to the simpler forms of agricultural industry. These, however, could hardly have arisen till after the invention of weapons and of tools, as well as the discovery of the use of fire, and it is by no means easy to see how natural selection alone, which can only produce modifications in accordance with an animal's needs, never beyond them, could have led to that mental superiority which at once placed man so far above all other animals, and have endowed him with such capacities for advancement as he actually possessed.

## CONCLUSIONS AS TO THE DEVELOPMENT OF HUMAN CHARACTER.

From the sketch which has now been given of the actual powers of the human mind, and of the various influences which may conceivably have modified it, we have been led to some very startling conclusions. We see, first, that the general idea that our enormous advances in science and command over nature serve as demonstrations of our mental superiority to the men of earlier ages, is totally unfounded. The evidence of history and of the earliest monuments alike go to indicate that our intellectual and moral nature has not advanced in any perceptible degree. In the second place, we find that the supposed great mental inferiority of savages is equally unfounded. The more they are sympathetically studied, the more they are found to resemble ourselves in their inherent intellectual powers. Even the so-long-despised Australians--almost the lowest in material progress--yet show by their complex language, their elaborate social regulations, and often by an innate nobility of character, indications of a very similar inner nature to our own. If they possess fewer philosophers and moralists, they are also free from so large a proportion of unbalanced minds--idiots and lunatics--as we possess. On the other hand, we find in the higher Pacific types, men who, though savages as regards material progress, are yet generally admitted to be--physically, intellectually, and morally--our equals, if not our superiors. These we are rapidly exterminating through the effect of *our* boasted civilisation!

Thirdly, we have no proof whatever that even the men of the stone age were mentally or morally inferior to ourselves. The case of the Pacific Islanders shows that simple arts and constructions with the absence of written language affords no proof of inferiority; while the undoubted absence of any selective power of "survival value" adequate to the evolution of the higher intellectual, æsthetic, and moral faculties-- which we find so fully developed in Ancient India, Egypt, and Greece--indicates that the very earliest men of whose existence we have any certain knowledge must also have possessed these faculties. If they did not possess them there must have been progressive mental progress independent of selection and without any intelligible cause.

One other characteristic of man which supports this conclusion is, as already shown, the extreme variability of his whole mental and moral nature, a variability much greater than that present in his body; and this again indicates that there has been no selective agency adequate to limit its range or guide it in any special direction.

## CONCLUSIONS AS TO THE ESSENTIAL NATURE OF HUMAN CHARACTER.

The preceding considerations lead us to conclude that the higher mental or spiritual nature of man is not the mere animal nature advanced through survival of the fittest. All the greatest writers and thinkers on the subject now admit this. In the last chapter of my *Darwinism* I have shown that some of the bodily characteristics of man are similarly inexplicable as the result of the same selective process. Darwin himself declared that the law of natural selection was, in his opinion, the greatest but not the exclusive means of modification.

To me it appears that, just as gravitation rules the whole material universe, so natural selection rules, and has ruled, the whole organic world. But in the countless modifications of matter, other quite distinct forces control or antagonise gravitation. Molecular and chemical forces, within their sphere of action, are far more powerful, and entirely neutralise the effect of the more far-reaching agency. Electricity and other ethereal forces are still more powerful; and, as seen at work in cometary emanations, oppose and overcome the gravitative force of the sun.

In like manner we see in the organic world a new and higher series of powers at work. First, in the life-force that renders possible the whole marvellous structure, growth, and products of the vegetable kingdom; next, in the higher life of consciousness and purposive action, as manifested in the lower animals; and, lastly, in the still higher spiritual nature--a little lower than that of the angels--with which man is endowed.

There is no evidence whatever that any of the animal forms below man possess the germs of this higher nature, however intelligent and teachable many of them undoubtedly are. Had they possessed it, some of them would have given indications of it. Such very diverse animals as the cat and the dog, the horse and the elephant, the monkey and the chimpanzee, exhibit a nearly equal amount of animal intelligence, but none of them can be said to be decidedly superior to the rest, none show any clear signs of the possession of even the rudiments of those faculties which raise man so infinitely above them all, or even of those much lower yet still essential powers, which enabled man, as soon as he became man, to develop language, to utilise fire, to make tools and weapons, to sow seeds, and to become shepherd and herdsman.

But from the epoch when man first attained to his specially human powers, he not only at once assumed command over the earth and all its forms of life, but commenced that development of latent faculties of which we find such striking evidences throughout history. The whole universe, in all its myriad forms, in all its intricacies of structure and motion, in all its marvellous beauty and inexhaustible utilities, in all its complex and mysterious laws and forces, became to him a vast school-room, furnishing the materials needed for the development of all his hitherto unused faculties and for the gradual elevation of his intellectual and moral nature. But the possibilities of such development must have pre-existed; the germs must have been present; every faculty must have been latent, or no amount of marvel and mystery could have called them forth.

How this higher nature originated in man, we may never know; but all the evidence points in the direction of some spiritual influx analogous to that which first initiated the organised life of the plant; then the consciousness and intelligence of the animal; and, lastly, reason, the sense of beauty, the love of justice, the passion for truth, the aspiration towards a higher life which everywhere, though in varying degrees, characterise the Human Race.

# THE POSSIBLE IMPROVEMENT OF CHARACTER.

Although, as we have seen, there has been no general advance of character during the whole period of which we can obtain any definite information, due to the absence of any great or constant selective agency, there is every reason to believe that it will be so improved in the not distant future. The heights to which it has attained in a few rare examples in all ages, taken in connection with the enormous range of variation it presents at the present time, show us that ample materials exist for raising its present average almost indefinitely. This can be effected by two distinct influences, which can and must always work together--education and selection by marriage. As yet we have no true and effective education. The very first essential in the teacher--true love of, or any sympathy with, the children--is not made one of the conditions of entering that great profession. Till this is made the *primary* qualification (as it was by Robert Owen at his schools in New Lanark) no real improvement in social and moral character can be effected. Mere intellectual instruction--which is all now given--is *not* a complete education, is really the least important half of it.

The other and more permanently effective agency, selection through marriage, will come into operation only when a greatly improved social system renders all our women economically and socially free to choose; while a rational and complete education will have taught them the importance of their choice both to themselves and to humanity. This subject I have treated in my *Studies* (see the chapter on "Human Selection"). It will act through the agency of well-known facts and principles of human nature, leading to a continuous reduction of the lower types in each successive generation, and it is the only mode yet suggested which will automatically and naturally effect this.

When we consider the enormous importance of such a continuous improvement in the average character, and that our widespread and costly religious and educational agencies have, so far, made not the slightest advance towards it, we shall, perhaps, realise, before it is too late, that we have begun at the wrong end. Improvement of social conditions must precede improvement of character; and only when we have so reorganised society as to abolish the cruel and debasing struggle for existence and for wealth that now prevails, shall we be enabled to liberate those beneficent natural forces which alone can elevate character.

The great lesson taught us by this brief exposition of the phenomena of character in relation to the known laws of organic evolution is this: that our imperfect human nature, with its almost infinite possibilities of good and evil, can only make a systematic advance through the thoroughly sympathetic and ethical training of every child from infancy upwards, combined with that perfect freedom of choice in marriage which will only be possible when all are economically equal, and no question of social rank or material advantage can have the slightest influence in determining that choice.

When our workers, our thinkers, our legislators can be persuaded to accept these fundamental truths, and make them the twin guiding stars of their aspirations and their efforts, the onward march towards true civilisation will have begun, and for the first time in the history of mankind, his Character--his very Human Nature itself--will be improved by the slow but certain action of a pure and beautiful form of selection--a selection which will act, not through struggle and death, but through brotherhood and love.

*Notes Appearing in the Original Work*

1. In the author's work, "Man's Place in the Universe," the various lines of evidence leading to this conclusion have been fully set forth.

2. *Natural Selection and Tropical Nature*, p. 181.